YOUR KNOWLEDGE HAS VALUE

- We will publish your bachelor's and master's thesis, essays and papers

- Your own eBook and book - sold worldwide in all relevant shops

- Earn money with each sale

Upload your text at www.GRIN.com and publish for free

Bibliographic information published by the German National Library:

The German National Library lists this publication in the National Bibliography; detailed bibliographic data are available on the Internet at http://dnb.dnb.de .

Imprint:

Copyright © 2018 GRIN Verlag
Print and binding: Books on Demand GmbH, Norderstedt Germany
ISBN: 9783668628960

This book at GRIN:

https://www.grin.com/document/388407

Patrick Kimuyu

Comparison of Velocity and Ultrasound Transit Time Spectroscopy in Cancellous Bone Phantom

GRIN Verlag

Comparison of Velocity and Ultrasound Transit Time Spectroscopy (UTTS) in Cancellous Bone Phantom

Name: Patrick Kimuyu

Introduction

Medical imaging technology plays an important role of creating internal images of the human body for clinical or medical purposes. Historically, this technology was born in November 1895 when Wilhelm Roentgen discovered electromagnetic radiation (x-ray) (Levine, 2010). Medical imaging technique can be defined as a technique which each modality could provide unique details of the human body function. The discovery of x-ray was a motivation reason for others to improve various technologies in medical imaging over the past years such as computed tomography (CT), ultrasound and magnetic resonance imaging (MRI) (Bradley, 2008). Ultrasound is one of the medical imaging technologies that are known as sound waves with a frequency above 20 KHz that excess the human hearing range using non-ionizing radiation. Ultrasound is a diagnostic modality technique that has been in clinical use over the past 40 years when Theodore Dussik and his brother Friederich in 1940s attempted to diagnose brain tumours using ultrasound waves, although their incredible work achieved success in 1970s (Newman & Rozycki, 1998). The aim of this study is to test the hypothesis that the minimum ultrasound transit time above noise (derived from the transit time spectrum) through cancellous bone may predict the velocity measurement. Therefore, deconvolution method has been used to predict ultrasound transit time through cancellous bone and then compare it to the reported transit time from clinical ultrasound bone densitometer (CUBA).

Overview

Ultrasound is a technique that involves creating an image from the returning waves after entering sound waves to the human body. The advantage of using ultrasound waves that are safe and not able to change cell structure is because; it has low energy which is used to visualize different areas of the body. This procedure does not cause damage to the structure of the cell, unlike x-ray which is a kind of ionization radiation that can change or damage the cell structure (Bradley, 2008).

Ultrasound image can be obtained by using a device called a transducer which works as an energy converter. This device converts electrical energy into mechanical energy, and there are techniques to measure ultrasound waves by using transducer device which known as pulse-eco and transmission techniques (Rizzatto, 1998).

The produced ultrasound image can describe many features of the artefacts and objects such as reflection, refraction, scattering and absorption due to the physical properties of the ultrasonic beam when interacts with tissue. Also, there are important quantities that can be used to define the produced ultrasound image such as the propagation speed, frequency, angle of incident, pulsed ultrasound and attenuation (Aldrich, 2007).

Ultrasound has been used widely over the past years for the assessment of bone due to the results of the speed of sound and attenuation of the sound wave that provide information of elasticity, density and structure of bone. Cancellous bone is one of the bones in the human body that have been used in this project for the assessment of osteoporosis (Rho, 1996). Osteoporosis can be defined as a disease which is manifested by a decrease of bone mass and density that lead to an increased risk of fracture (Sugerman, 2014). Different methods have been used to determine bone mineral density (BMD) including the dual energy x-ray absorptiometry (DEXA), quantitative x-ray computed tomography (QCT) and quantitative ultrasound (QUS) in the process of diagnosing osteoporosis (Shan et al., 2013).

There are different kinds of machines that are used to determine ultrasound waves such as Achilles (GE Medical Lunar), Quidel QUS-2 (Biomedx) and McCue CUBA Clinical. In this project, CUBA clinical system has been used to determine ultrasound waves. Some of the most significant properties of using CUBA clinical system are that the portable device has automatic transducers to be in direct contact to the sample, calf support and area to insert sample for positioning improvement (Langton et al., 1990).

CUBA clinical system is designed to measure time of flight (TOF), also called ultrasound transit time (UTT) and broadband ultrasound attenuation (BUA) can be used for the assessment of the fracture risk. Time of flight (TOF) is defined as the travelled time for ultrasound wave that propagates through a medium. However, velocity and BUA measurements are important to determine osteoporosis where the known information of the sample thickness and transit time is used to estimate the velocity of ultrasound. This helps in predicting the density and elasticity of bone, whereas the known information of broadband ultrasound attenuation helps in predicting the density and structure of the cancellous bone (Langton, Wille & Flegg, 2014).

Background

The Nature of Ultrasound

Ultrasound is a sound or pressure wave that has a frequency of more than 20 KHz; this frequency is higher than the one detected by the human ear (Bertora, 2007). Ordinarily, ultrasound propagates as longitude waves through fluid, air and human tissue due to the changes in pressure with slow speed of propagating in all materials such as soft tissue; about 1540 m/s. The number of cycles or pressure changes in 1 second, known as the frequency of ultrasound, can be determined by the sound source only. It cannot be determined by the medium of the travelling sound, and the range of frequencies used in clinical procedures is between 2 to 10 MHz. The speed of sound waves that travel through a medium can be determined by the density and stiffness of that medium as it is demonstrated in figure 1. The changes in either density or stiffness will affect the pulse transit time where pulsed beams are used in clinical procedures to achieve the required resolution (Aldrich, 2007).

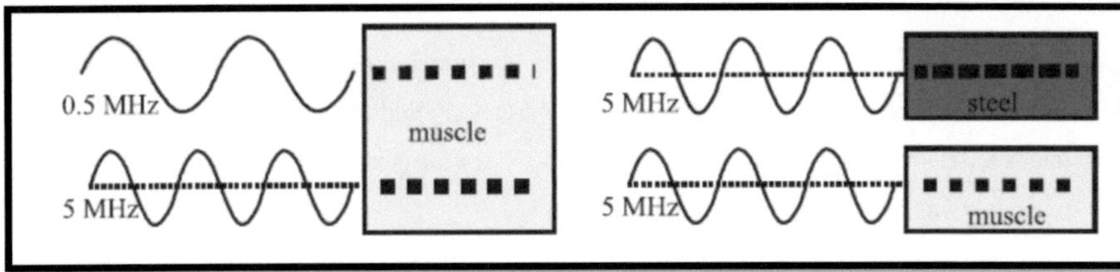

Figure 1: Speed of ultrasound in different materials (Aldrich, 2007 p. S132)

Production of Ultrasound

Transducer is a small device that is responsible for producing ultrasound waves by receiving electrical energy from the source and converts it to mechanical vibrations (transit mode). The mechanical vibrations are, in turn, converted into electrical energy (receive mode) that works on a piezoelectric effect (Rizzatto, 1998). Piezoelectric materials have the ability to produce an electrical field when they are mechanically deformed, but in terms of applying an electrical field as a voltage pulse to piezoelectric materials, then the deformation will occur to the materials. Ceramic materials such as Lead titanate, Lead zirconate titanate (PZT) and Barium titanate are the standard piezoelectric materials that have been used for medical imaging processes (Whittingham, 2007).

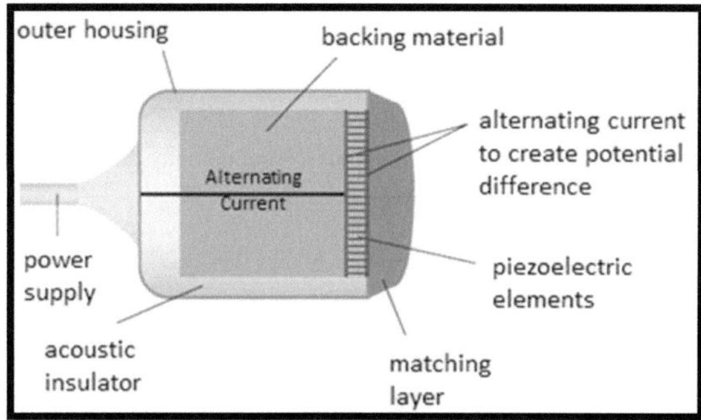

Figure 2: Transducer Basics (Agnieszka, n.d.)

Ultrasound Measurement Techniques

There are two techniques used to determine the measurement of ultrasound; the first one is pulse-eco technique and the second is transmission technique. Pulse-eco technique uses a single transducer to transmit and receive signal when the reflected signal from the area of interest detected by the same transducer which transmit that signal to propagate through this area (Bertora, 2007). However, the transmission technique used in this project uses two transducers, one of them acts as a transmitter of the signal and the other as the receiver of the signal that propagates through a medium (Center, 2001). A large amount of piezoelectric will be used in transmission technique, and it is preferred to be used for application to bone due to the high attenuation of bone nature (Resnick, 1988).

Ultrasound Interactions with Tissue

Ultrasonic waves have been used for diagnostic and therapeutic purposes in medical imaging processes (Droin, Berger & Laugier, 1998). Acoustic impedance properties of matter can be used to determine the interactions of ultrasound while propagating the ultrasonic energy into a selected area of the tissue. In an ultrasound, attenuation, reflection, refraction, scattering, absorption and interference are types of the interactions that will occur (Aldrich, 2007).

Attenuation

Attenuation occurs when the ultrasound waves loss their energy while travelling through tissues that result to reduce the intensity of the travelled beam. The loss of intensity of the ultrasound beam happens due to several processes such as refraction, absorption, scattering and reflection. As frequency increases, the attenuation increase in a given materials (Bestan, 2012).

Reflection

Reflection of ultrasound is the most important process in producing an ultrasound image that depends on acoustic impedance and acoustic boundaries (tissue interfaces). Acoustic impedance of the material (Z) is defined as the density of the medium (ρ) and velocity of ultrasound of that medium (v).

Z = velocity (v) x density (ρ)

Acoustic boundaries (tissue interfaces) known as the position of the tissue that the values of the acoustic impedance are changed, and it is important for the interaction of ultrasound (Center, 2001). Reflection of ultrasound occurs at boundaries of tissue when sound waves enter between two different tissues, some of it is reflected back to the medium and the rest of it passes to the second medium. A strong reflection occurs if the difference of the acoustic impedance is large while total reflection is produced when the acoustic impedance differences is very large (Aldrich, 2007).

Scattering

Non-specular reflection or scattering is one of the ultrasound interaction processes with matter that depends on the interface shape and ultrasound beam dimensions. If the shape of the reflected boundary is irregular and the ultrasound beam has a diameter bigger than the dimensions of the reflected boundary, reflection occurs for the incident beam in several different directions which known as scattering of ultrasound. Ordinarily, scattering of ultrasound increases as the frequency increase where scattering illustrate strong frequency dependence (Nelligan, n.d.).

Refraction

Direction of beam at the boundary between two media is changes when ultrasound wave passes with different speed of sound. Frequency of the incident beam does not change where the ultrasonic waves pass through the first medium into the second with change of the wavelength (Aldrich, 2007).

Velocity = wavelength x frequency

Absorption

Absorption plays a crucial role of ultrasound attenuation. When sound waves travel through a propagating medium, heat is produced because of friction and loss of energy from the incident sound waves. Relaxation times of medium, viscosity of medium and beam frequency are the main variables that affect the absorption of ultrasound in a medium. Therefore, the bone absorbs ultrasound more than soft tissue where the intensity of sound in soft tissue decreases with depth (Humphrey, 2007).

Ultrasonic Wave Propagating In Bone

Ultrasound has been used for the assessment of bone defects where the speed of sound measurements through a medium such as cancellous bone provides evidence of bone elasticity and density. On the other hand, attenuation measurements provide evidence of the bone structure (Turner & Eich, 1991). Cancellous bone, commonly referred to as spongy bone consists of porous bone with large spaces filled with bone marrow, and the organized framework of that bone is called trabeculae. However, it is worth noting that, the results of using ultrasonic wave in human bone to investigate the properties of the bone and identify osteoporosis have become a significant issue for several past years (Langton, Wille & Flegg, 2014).

Osteoporosis in Cancellous Bone

A normal bone consists of a mixture of minerals such as calcium; protein, phosphate and magnesium. Therefore, the loss of bone minerals refer to the loss of bone density which is one of the imaging techniques called bone mineral density (BMD) used to identify osteoporosis. Osteoporosis is defined as a disease which is manifested by a decrease of bone mass and density that lead to an increased risk of fracture (Osteoporosis, 2012). Figure 3 shows the difference between the normal bone (a) and osteoporosis bone (b).

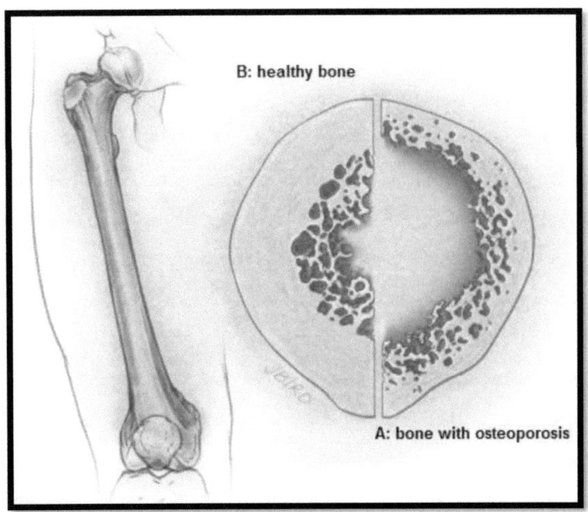

Figure 3: The difference between normal bone (a) and Osteoporosis bone (b) (Sugerman, 2014).

Methods of Determining Osteoporosis

There are different methods for determining BMD including dual energy x-ray absorptiometry (DEXA), quantitative x-ray computed tomography (QCT) and quantitative ultrasound (QUS).

DEXA is a common technique for measuring BMD and diagnosis of osteoporosis due to the absorption of calcium by x-rays. DEXA can be used to determine bone density in hip and spine by using two different beams of x-ray which pass through the bone and soft tissue. It also determines the bone density by comparing the amount of blocked x-rays. DEXA technique uses low dose radiation, fast and it can measure 2% of bone loss in one year (Osteoporosis, 2012).

On the other hand, QCT measures BMD and the advantage of using this technique is measuring BMD in three dimensional structure of hip and lumbar spine compared to DEXA technique (Wakamatsu, Matsuyama, Yabana, 1998). QCT technique is also suitable for obesity patients who are suffering from spinal degenerative diseases or disc space narrowing because it provides better resolution compared to DEXA (Shan et al., 2013).

QUS is a non-invasive and radiation free technique compared to x-ray. It is a reliable technique for the assessment of osteoporosis diagnosis and bone status because; it can estimate velocity and absorption from the human bone. BUA and Speed of sound (SOS) are the main parameters of QUS that are used for bone diagnoses. The advantage of using this technique is that it is not expensive, and it provides information about bone density more than other techniques. In addition, the device is portable; thus, it is easy to use (Glüer, 1997).

Ultrasound Densitometry

The use of ultrasound is to define the assessment of bone based on the attenuation of sound waves and the speed of sound that has been affected by density, elasticity and structure of the material. Measurements of ultrasound velocity can predict the density and elasticity of bone, especially in calcaneus bone site due to some reasons compared to the measure in other sites of the bone. For example, vertebrae and hip bones are difficult to measure using ultrasonic waves because of the ultrasound signal attenuates due to the complexity of soft tissue surrounding these bones. However, there are reasons that calcaneus bone is preferred as the best site for measuring velocity and predicts elasticity and density. First, this bone consists of a flat surface which makes sure a good connection between the transducers and area of interest and it contains little soft tissue that makes it easy to measure. Secondly, calcaneus bone consists of approximately 90% of trabecular bone (Langton, 2011).

In terms of the attenuation of ultrasound, it can be determined when the signal of the ultrasound wave passes through a medium with a known thickness. The structure of cancellous bone and density are also related to the measurements of ultrasound attenuation where the ultrasonic attenuation increases with frequency (Langton & Wille, 2013). BUA measurements help to provide information about the osteoporotic bone fracture risk. This is determined by comparing the spectrum of the amplitude received signal to the spectrum of a reference medium and the difference between them is plotted against frequency where the slope of liner is obtained and defined as BUA index (dB/MHz). In practice, dividing the BUA index by the thickness of the sample helps in determining the volumetric parameter (dB/MHz.cm) (Langton et al., 1990).

Previous Studies

Measurements of broadband ultrasound attenuation (BUA) and the speed of sound (SOS) have been studied widely in both vitro and vivo for the assessment of cancellous bone.

In 1990, Langton used a clinical ultrasound bone densitometer system (CUBA) to determine the speed of sound and broadband ultrasound attenuation measurements on cancellous and cortical bone. The properties of using CUBA clinical system were rapid, portability and non-invasive technique that was performed without using a tank of water as what have been done in the previous studies of measuring BUA. A transducer was in direct contact with the area of interest by using a coupling gel and the measurements of overlying soft-tissue determined an exact measurement to obtain bone density. Also, attenuation data were normalised to a volumetric density parameter. Therefore, CUBA clinical system was found to be an accurate and it has potential as a method for the clinical population at risk of fracture (Langton et al., 1990).

Later in 1999, Trebacz and Natali examined the influence of micro-architecture and density on ultrasound wave that was propagated through cancellous bone with twenty different samples of vertebra and calcanei. Lateromedial in calcaneus and anteroposterior in vertebra were the directions of bones which ultrasound wave passed through to investigate the relationship between density, velocity, and BUA of ultrasound in the bone sample. Therefore, positive correlations occurred between BUA and velocity of ultrasound with morphometric and density for calcaneus and vertebra where the poorest correlation in calcaneus ($r = 0.632$, $p = 0.002$) was found between the thickness of trabecular and velocity. On the other hand, the finest correlation ($r = 0.961$, $p = 0.0001$) was found between velocity of ultrasound and the sample of bone density in vertebra. In terms of BUA, the finest correlation was with trabecular bone volume and trabecular thickness in vertebra ($r = 0.961$, $p = 0.0001$) (Trebacz & Natali, 1999).

The Biot's theory was developed in 1956, when this theory was used in the context to test the geophysical characteristics of fluid-saturated porous rocks to predict the acoustical properties by using ultrasound wave. Therefore, it is apparent that, ultrasound waves propagate through cancellous bone that consist of solid elastic structure (bone) and interspersed fluid (marrow). Various parameters were found to be measured that affect the propagation of ultrasound to be predicted. However, Biot's theory is used to predict the velocity of ultrasound in several studies by different authors, but the prediction of ultrasound attenuation remain indefinable (McKelvie & Palmer, 1991).

References

Agnieszka (n.d.). *Ultrasound Physics E-Portoflio, Artifacts-5.* Retrieved from https://sites.google.com/site/agnieszkasphysicseportfolio/artifact-5

Aldrich, J. E. (2007). Basic Physics of Ultrasound Imaging. *Crit Care Med.,* 35(5): S131-7. Retrieved from http://online.uminho.pt/pessoas/smcn/MAIO/ultra-som/ultrasound%20physics%20Aldrich%202007.pdf

Bertora, F. (2007). *Ultrasound Transducers. In Y. Lemoigne, A. Caner & G. Rahal. (Eds) Physics for Medical Imaging Applications (p. 111-121).* Dordrecht, Netherlands: Springer.

Bestan, J. (2012). *Basic Ultrasound Physics.* Retrieved from: http://www.wikiradiography.com/page/Basic+Ultrasound+Physics

Bradley, W.G. (2008). History of Medical Imaging 1. *Proceedings of the American Philosophical Society,* 152(3): 349-361.

Center, N.E. (2001). *Reflection and Transmission Coefficients.* Retrieved from: http://www.ndt-ed.org/EducationResources/CommunityCollege/Ultrasonics/Physics/reflectiontransmission.htm

Droin, P., Berger, G. & Laugier, P. (1998). Velocity Dispersion of Acoustic Waves in Cancellous Bone. *IEEE Trans Ultrason Ferroelec Freq Contr.,* 45(3): 581-592.

Glüer, C. C. (1997). Quantitative Ultrasound Techniques for the Assessment of Osteoporosis: Expert Agreement on Current Status. *Journal of Bone and Mineral Research,* 12(8): 1280-1288.

Humphrey, V. F. (2007). Ultrasound and Matter: Physical Interactions. *Progress in Biophysics and Molecular Biology,* 93(1–3): 195-211.

Langton, C. et al. (1990). A Contact Method for the Assessment of Ultrasonic Velocity and Broadband Attenuation in Cortical and Cancellous Bone. *Clinical Physics and Physiological Measurement,* 11(3): 243.

Langton, C. M. & Wille, M. L. (2013). Experimental and Computer Simulation Validation of Ultrasound Phase Interference Created by Lateral in Homogeneity of Transit Time in Replica Bone: Marrow Composite Models. Proceedings of the Institution of Mechanical Engineers, Part H. *Journal of Engineering in Medicine,* 227(8): 890-895.

Langton, C. M. (2011). The 25th Anniversary of BUA for the Assessment of Osteoporosis: Time for a New Paradigm? Proceedings of the Institution of Mechanical Engineers, Part H. *Journal of Engineering in Medicine*, 225(2): 113-125.

Langton, C. M., Wille, M. L., & Flegg, M. B. (2014). A Deconvolution Method for Deriving the Transit Time Spectrum for Ultrasound Propagation through Cancellous Bone Replica Models. *Proc Inst Mech Eng H*, 228(4):321-9. doi: 10.1177/0954411914523582

Levine, H. (2010). *Medical Imaging*. Santa Barbara, CA: ABC-CLIO.

McKelvie, M. & Palmer, S. (1991). The Interaction of Ultrasound with Cancellous Bone. Physics in Medicine and Biology, 36(10): 1331.

Nelligan, T. (n.d.). Ultrasonic Material Analysis. Retrieved from: http://www.olympus-ims.com/en/applications-and-solutions/introductory-ultrasonics/introduction-material-analysis/

Newman, P.G. & Rozycki, G.S. (1998). The History of Ultrasound. *Surgical Clinics of North America*, 78(2): 179-195.

Resnick, H. A. (1988). *Piezoelectricity Effect*. Retrieved from: http://hyperphysics.phy-astr.gsu.edu/hbase/solids/piezo.html

Rho, J. Y. (1996). An Ultrasonic Method for Measuring the Elastic Properties of Human Tibial Cortical and Cancellous Bone. *Ultrasonics*, 34(8): 777-783.

Rizzatto, G. (1998). Ultrasound Transducers. *European Journal of Radiology*, 2: S188-S195.

Shan, P. F. et al. (2013). Osteoporosis. *International Journal of Endocrinology*, 2013(2013): 2.

Sugerman, D. (2014). Osteoporosis. *JAMA*, 311(1): 104-104.

Trebacz, H. & Natali, A. (1999). Ultrasound Velocity and Attenuation in Cancellous Bone Samples from Lumbar Vertebra and Calcaneus. *Osteoporosis International*, 9(2): 99-105.

Turner, C. H. & Eich, M. (1991). Ultrasonic Velocity as a Predictor of Strength in Bovine Cancellous Bone. *Calcified Tissue International*, 49(2): 116-119.

Wakamatsu, H., Matsuyama, T. & Yabana, T. (1998). [Quantitative X-ray computed tomography]. Nihon rinsho. *Japanese Journal of Clinical Medicine*, 56(6): 1479-1483.

Whittingham, T. A. (2007). Medical Diagnostic Applications and Sources. *Progress in Biophysics and Molecular Biology*, 93(1–3): 84-110.

YOUR KNOWLEDGE HAS VALUE

- We will publish your bachelor's and master's thesis, essays and papers

- Your own eBook and book - sold worldwide in all relevant shops

- Earn money with each sale

Upload your text at www.GRIN.com and publish for free